A LOOK AT LIFE S

WHAT ARE PLANTS?

BY KATE MIKOLEY

Gareth Stevens PUBLISHING

CRASHCOURSE

Please visit our website, www.garethstevens.com. For a free color catalog of all our high-quality books, call toll free 1-800-542-2595 or fax 1-877-542-2596.

Cataloging-in-Publication Data

Names: Mikoley, Kate.
Title: What are plants? / Kate Mikoley.
Description: New York : Gareth Stevens Publishing, 2020. | Series: A look at life science | Includes glossary and index.
Identifiers: ISBN 9781538248591 (pbk.) | ISBN 9781538248614 (library bound) | ISBN 9781538248607 (6 pack)
Subjects: LCSH: Plants--Juvenile literature. | Plant physiology--Juvenile literature. | Plant ecology--Juvenile literature.
Classification: LCC QK49.M53 2020 | DDC 580--dc23
First Edition

Published in 2020 by
Gareth Stevens Publishing
111 East 14th Street, Suite 349
New York, NY 10003

Designer: Reann Nye
Editor: Kate Mikoley

Photo credits: Series art MriMan/Shutterstock.com; cover, p. 1 AustralianCamera/Shutterstock.com; p. 5 Jakkrit Orrasri/Shutterstock.com; p. 7 hovikphotography/Shutterstock.com; p. 9 Perfect Lazybones/Shutterstock.com; p. 11 stockphoto mania/Shutterstock.com; p. 13 krasky/Shutterstock.com; p. 15 Agatha Koroglu/Shutterstock.com; p. 17 Romas_Photo/Shutterstock.com; p. 19 lovelyday12/Shutterstock.com; p. 21 TuyNuy/Shutterstock.com; p. 23 nito/Shutterstock.com; p. 25 Lazartivan/Shutterstock.com; p. 27 Adam J/Shutterstock.com; p. 29 Shannon West/Shutterstock.com; p. 30 (plant) adike/Shutterstock.com; p. 30 (sun) stockelements/Shutterstock.com; p. 30 (sugar) Saranya Loisamutr/Shutterstock.com; p. 30 (clouds) KC Lens and Footage/Shutterstock.com; p. 30 (water) Only background/Shutterstock.com.

Printed in the United States of America

CONTENTS

Words in the glossary appear in **bold** type the first time they are used in the text.

PLANTS ALL AROUND

Plants can be found all over the world! Some grow on land, while others live in water. These wonderful **organisms** are very important for life on Earth. In fact, humans and other animals wouldn't be able to live without them!

MAKE THE GRADE

All **food chains** include plants. Even animals that don't eat plants get **nutrients** because of animals that did eat plants.

ALL KINDS OF PLANTS

There are thought to be more than 390,000 different species, or kinds, of plants. Some are so small they can only be seen with special tools. Others, such as giant sequoia trees, can be taller than 300 feet (91 m)!

MAKE THE GRADE

One giant sequoia in California measures more than 100 feet (30 m) around at its base.

FLOWER POWER

Most plants are flowering plants. These are plants that produce parts where the seed or fruit grows from, called flowers. This includes plants we commonly think of as flowers, such as roses. However, other plants, such as many trees and grasses, are also flowering plants.

MAKE THE GRADE

Flowering plants can live in many types of **environments**, but most kinds live in hot and wet areas near the **equator**.

MAKING MORE PLANTS

Flowering plants reproduce, or make new plants, using pollination. Part of the flower called the anther produces pollen. For a seed to be made, the pollen needs to make its way to another flower part, called the stigma. Pollen is often moved by insects.

MAKE THE GRADE

Some flowers have both anthers and stigmas. Others have one or the other, so pollen needs to move to another flower in order for pollination to happen.

Not all plants have flowers. Some, such as pine trees, still use seeds to reproduce, but their seeds are in cones rather than flowers. Other nonflowering plants don't use seeds at all. Instead, they use **spores** to reproduce.

MAKE THE GRADE

Ferns and mosses are examples of plants that use spores to reproduce.

GERMINATION STATION

When a plant grows from a seed it's called germination. Inside the seed is the embryo, or the beginning of the new plant. The seed also has food for the plant. When planted in the right environment, the seed will germinate.

MAKE THE GRADE

Seeds are often planted by people. Other times, they end up in the ground in more natural ways, such as being blown by the wind or moved by animals.

Different plants need different **conditions** in order to grow. All seeds need **oxygen** and water. They also need the right **temperature** and the right amount of light. When the conditions are just right, the seed takes in the oxygen and water.

MAKE THE GRADE

Some seeds need darkness to grow, while others need light. Likewise, some plants need different temperatures to grow.

The embryo within the seed begins to get bigger. Soon, the outer part of the seed breaks open. The plant's roots begin to grow into the ground. The roots give the plant a strong base and help keep it in place.

MAKE THE GRADE

After the roots form, the plant breaks through the ground so it can be seen. This part of the plant is called the shoot.

MORE GROWTH

From the shoot, the plant can continue growing a stem and leaves. Over time, the plant may produce flowers, where the seeds are formed. Or, it may start to make spores or cones. This starts the **life cycle** over again for new plants!

MAKE THE GRADE

In some plants, each flower only makes one seed. Other plants produce hundreds of seeds per flower.

The roots of the plant take in water and nutrients, which the plant needs to live. These travel through the stem to get to the leaves. The stem also lifts the plant toward the sun. The plant needs sunlight to live, too.

MAKE THE GRADE

Leaves are connected to the stem by a narrow stem called a petiole.

MAKING FOOD

Green plants make their own food through photosynthesis. During photosynthesis, the plant takes in **carbon dioxide** and water. It also takes in light from the sun. Using **energy** from this sunlight, it turns the carbon dioxide and water into a sugar called glucose.

MAKE THE GRADE

The glucose the plant makes is broken down and turned into energy. This energy helps the plant grow and stay healthy.

During photosynthesis, plants also make oxygen. This oxygen is then released, or let out, into the air for humans and other animals to breathe. Nearly all the oxygen in our world comes from photosynthesis. Without it, most organisms wouldn't be able to live!

MAKE THE GRADE

Plants that went through photosynthesis millions of years ago helped form the **fossil fuels** that we use today, such as coal.

PLANT YOUR OWN!

Scientists are still discovering and studying new plants today. As you get older, more and more facts about plants will become known. If you want to learn more about plants, ask an adult to help you plant some of your own!

MAKE THE GRADE

Caring for your own plants can teach you more about them. You don't need a big garden. Some plants do well in small pots indoors!

PHOTOSYNTHESIS

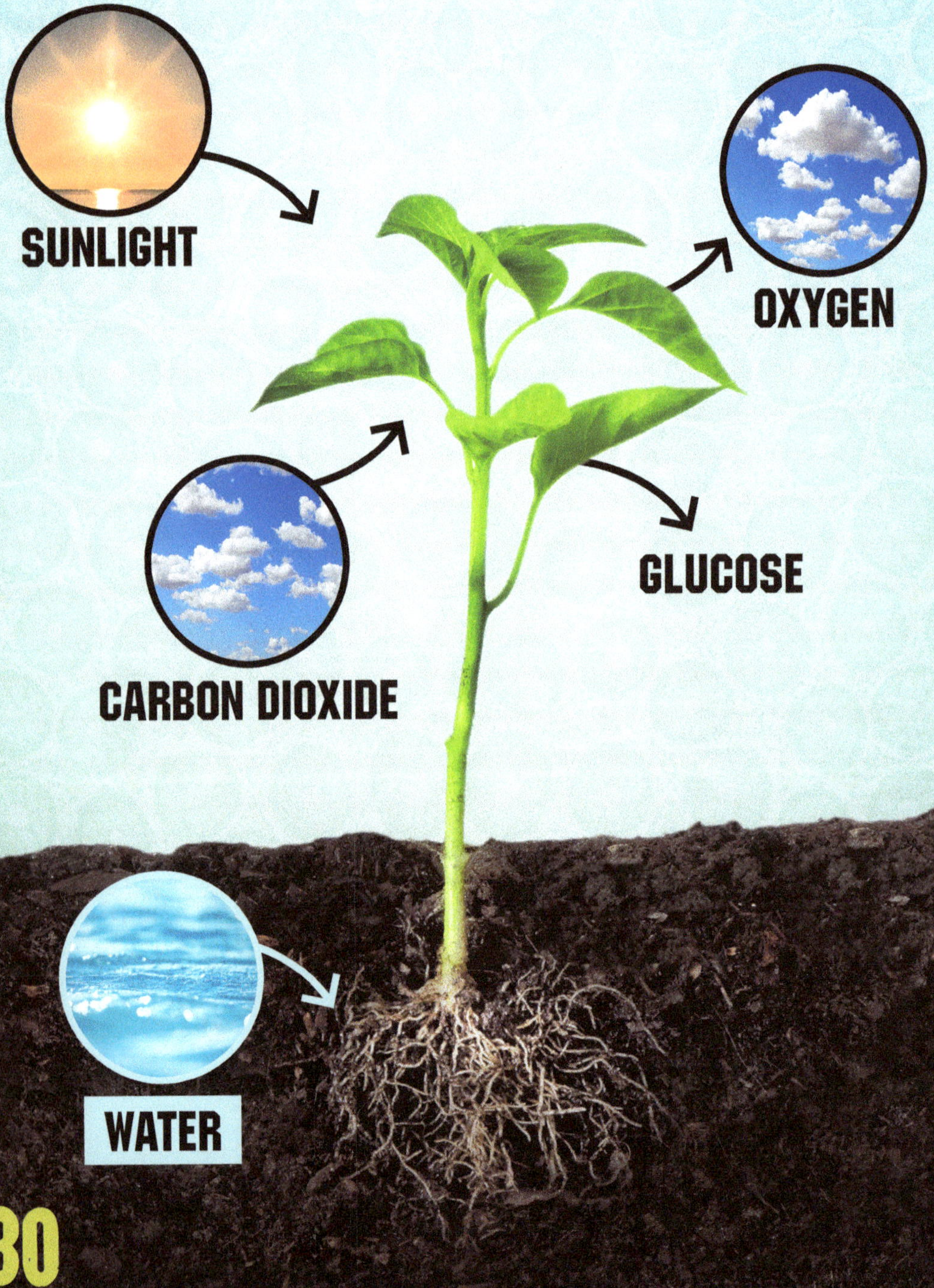

GLOSSARY

carbon dioxide: a gas breathed out by animals

conditions: the state in which someone or something lives

energy: power used to do things

environment: the conditions that surround a living thing and affect the way it lives

equator: an imaginary line around Earth that is the same distance from the North and South Poles

food chain: a series of organisms in which one uses the next lowest one as food

fossil fuel: matter formed over millions of years from plant and animal remains that is burned for power

life cycle: the stages through which a living thing passes in its life

nutrient: something a living thing needs to grow and stay alive

organism: a living thing

oxygen: a colorless, odorless gas that many animals, including people, need to breathe

spore: a small body made by a plant that can grow into another plant

temperature: how hot or cold something is

FOR MORE INFORMATION

BOOKS

Duling, Kaitlyn. *A Plant That Eats Spiders and Other Cool Green-and-Growing Facts.* North Mankato, MN: Capstone Press, 2019.

Gibbs, Maddie. *Types of Plants.* New York, NY: Britannica Educational Publishing, 2019.

WEBSITES

Plant a Garden
kids.nationalgeographic.com/explore/nature/plant-a-garden/
Find out how to grow some plants of your own on this page.

Plants
www.dkfindout.com/us/animals-and-nature/plants/
Learn more about many kinds of plants here.

Publisher's note to educators and parents: Our editors have carefully reviewed these websites to ensure that they are suitable for students. Many websites change frequently, however, and we cannot guarantee that a site's future contents will continue to meet our high standards of quality and educational value. Be advised that students should be closely supervised whenever they access the internet.

INDEX